Separations Division 2020

Held at the 2020 AIChE Spring Meeting and 16th Global Congress on Process Safety

Online
17 - 21 August 2020

ISBN: 978-1-7138-2340-7

Printed from e-media with permission by:

Curran Associates, Inc.
57 Morehouse Lane
Red Hook, NY 12571

Some format issues inherent in the e-media version may also appear in this print version.

AIChE
120 Wall Street, FL 23
New York, NY 10005-4020

Phone: (800) 242-4363
Fax: (203) 775-5177

www.aiche.org

Additional copies of this publication are available from:

Curran Associates, Inc.
57 Morehouse Lane
Red Hook, NY 12571 USA
Phone: 845-758-0400
Fax: 845-758-2633
Email: curran@proceedings.com
Web: www.proceedings.com

TABLE OF CONTENTS

590404 Practical Industrial Distillation

Abstract:

This presentation, aimed at young professionals in Separations, provides an overview of industrial distillation - from theory to practice.

590410 Distillation Trays and Packings

Abstract:

This paper provides a unique insight into the internals and operations of trayed and packed fractionation towers.

590405 Dividing Wall Distillation Columns

Abstract:

This paper presents an introduction to dividing wall columns and their applications in the chemical process industry

590298 Distillation Column Control

Abstract:

The understanding of the basic control schemes on a distillation column is key for effective operation and troubleshooting. This paper introduces distillation column control, and the concepts behind typical control schemes employed in industrial distillation columns. The advantages, disadvantages, and pitfalls of control configurations will be illustrated using examples.

590406 Introduction to Troubleshooting Distillation Columns

Abstract:

This presentation is aimed at providing beginner tips and techniques for troubleshooting distillation columns.

590409 Radioisotope Techniques for Process Investigations

Abstract:

In all chemical plants and refineries there generally is not enough instrumentation in the right places in
our processes. During the design phase, decisions were made between knowing more about the process,
which means buying more instruments, and saving capital funds. When a problem appears with the operation of a vessel, there are always gaps in the knowledge of what is happening due to a lack of instruments in the places where knowledge is not routinely required. Process engineers and operations personnel need to know what is happening so they can take corrective action. When knowledge is missing, they often take actions that make matters worse.

For distillation and separation processes Chemical Engineers have historically relied upon plant process
measurements such as flow rates, temperatures, and pressures and model or simulation results for data with which to solve troubleshooting or re-design problems. An array of on-line diagnostic services are available that can provide additional data offering real-time information on how pieces of process equipment are actually operating. The most common application is Gamma Scanning. This test is primarily applied, but not necessarily limited to, distillation or separation columns. Gamma scans provide a density profile of the internal process of operating distillation columns and other process vessels. The density profile can be used to diagnose the hydraulic operating conditions of mass transfer
devices such as damage to internals, flooding, degree of entrainment or weeping, liquid levels on trays and distributors, liquid distribution through packed beds, etc. The presenter will show case studies where scanning revealed vital process information that helped solve an operating problem, or helped make a revamp/re-design successful. Additional derivatives from gamma scanning of detecting flow distribution patterns will be demonstrated. The ThruVision Scan technique is a specialized horizontal gamma scan used to generate a topographic profile of the internal cross-sectional density of process equipment. This profile is useful for the detailed study of liquid flow distribution through packed columns as well as having applications beyond distillation columns. A hybrid application involving tracers
and scanning on fixed bed catalytic reactors will also be discussed.

AUTHOR INDEX

NOTES:

NOTES:

NOTES:

NOTES:

AIChE
120 Wall Street, FL 23
New York, NY 10005-4020

ISBN 978-1-7138-2340-7
9 781713 823407

CURRAN ASSOCIATES INC.
proceedings.com